Beautiful Planets
For Kids

Nature Books for Kids
By K. Bennett
Mendon Cottage Books

JD-Biz Publishing

**Download Free Books!
http://MendonCottageBooks.com**

All Rights Reserved.
No part of this publication may be reproduced in any form or by any means, including scanning, photocopying, or otherwise without prior written permission from JD-Biz Corp Copyright © 2016.

All Images Licensed by Fotolia and 123RF.

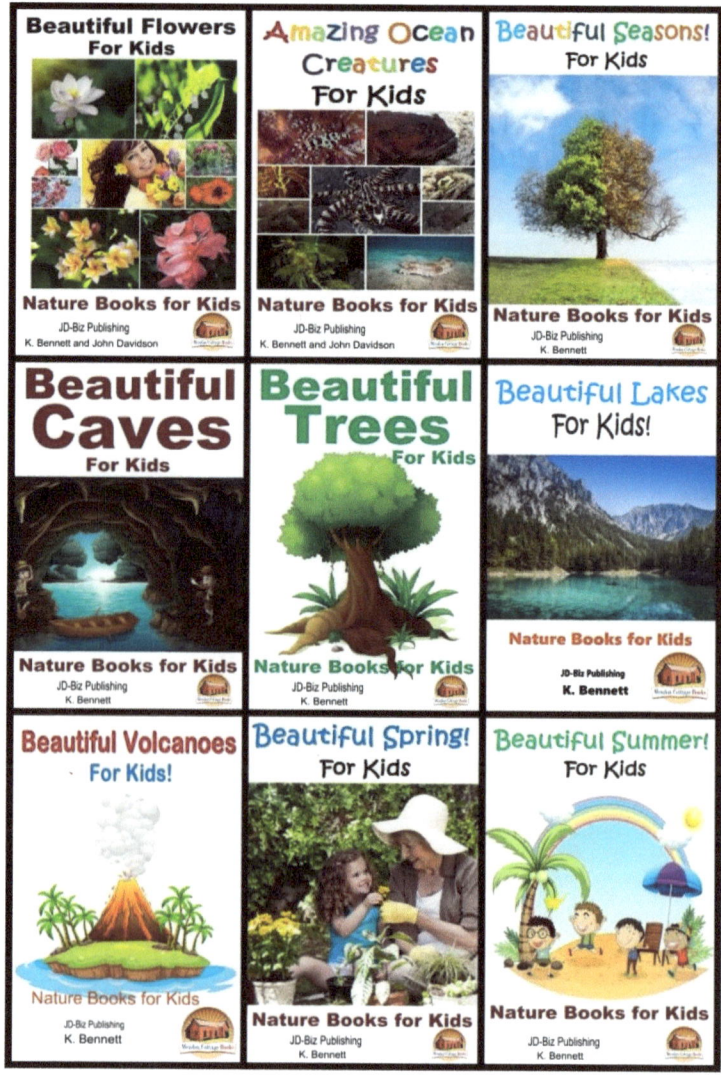

Download Free Books!
http://MendonCottageBooks.com

Table of Contents

Introduction 4

Chapter 1: The Sun 8

Chapter 2: Some Planet Basics 10

Chapter 3: Mercury 14

Chapter 4: Venus 16

Chapter 5: Earth 18

Chapter 6: Mars 20

Chapter 7: Jupiter 22

Chapter 8: Saturn 24

Chapter 9: Uranus 26

Chapter 10: Neptune 28

Chapter 11: Pluto 30

Chapter 12: Interesting Facts 32

Conclusion: 36

Sources: 39

Author Bio 40

Publisher 42

Introduction

Space, the final frontier… to explore strange new worlds, to seek out new life and new civilizations, to boldly go where no man has gone before. ~ **Gene Roddenberry**

We are living in an amazing place in the universe called: The Milky Way Galaxy. It is surrounded by lots and lots of stars, planets, asteroids, comets, and other celestial objects.

One neat place in the Milky Way Galaxy is where planet earth is found. Can you guess where we are? Did you guess: The solar system?

Good job!

The solar system has lots of fascinating things to discover. Let's learn about some of them and don't forget to share with others!

First, let's define our solar system. What is it? If someone asked you that question, what would you say?

ESA for kids explains it in a nice and simple way: "The Solar System is made up of the Sun and all of the smaller objects that move around it."

Simple enough, right? It might sound that way, but it isn't!

The solar system has eight planets. Let's start with the sun. It is the biggest part of our solar system and everything moves around this bright star.

Remember the name of the planets?

-Mercury
-Venus
-Earth

- Mars
- Jupiter
- Saturn
- Uranus
- Neptune

What about Pluto?

Don't worry...we will talk about this planet later.

There are other objects like asteroid belts. One of them is just after Mars and full of floating rocks. We can also find moons, gas, space dust, and comets.

I hope you enjoy this journey to explore places… where no one has gone before!

Chapter 1: The Sun

The sun is a big part of the solar system and extremely important to life on earth. It is not only hot enough to keep us warm, but it helps things to grow. Do you remember what plants need to stay alive?

Sunlight? Great job!

In our book <u>Beautiful Summer for Kids</u>, we explained: "Plants need the warm sun to grow…nice and strong." Sunlight is not only important for energy but also for food."

So why is sunlight so important? *Let's find out!*

The sun sits in the middle of the solar system, and the planets and objects race around it. I say race, because we are traveling very, very fast. Do you know how fast planet earth is traveling around the sun?

Scientists measured the distance and say around 67,000 miles per hour. If you prefer to use kilometers, that's 107,300 kilometers per hour!

Let's go back to the sun.

The sun "holds" the planets in place because it has a strong gravitation pull. This powerful force is called the "gravitational epicenter of the solar system." This means it keeps all the space bodies around it in orbit.

Why is this necessary?

Suppose Mars bumped into the earth? How about Jupiter? Do you think anything bad will happen?

This force to keep everything in place is so amazing that a comment on answers.com says, "No sun – no solar system."

Chapter 2: Some Planet Basics

Let's look at the planets in our solar system. They are divided into two different parts. One is called: Terrestrial planets and the other is called Jovian Planets. What makes them different?

Terrestrial planets: The four planets that are closest to the sun are inner planets or terrestrial planets. They get their names because they have a rocky surface like the Earth.

Jovian planets: The four planets that are farther away from the sun are known as outer planets or Jovian planets. Does the name sound strange to you? That's because they are: "Jupiter like" planets. They are made of gases and helium and do not have a rocky surface like the four inner planets.

Even the Jupiter like planets are different. Jupiter and Saturn are called gas giants. Uranus and Neptune are called ice giants. Can you guess why?

Simple…Jupiter and Saturn have lots more gas and Uranus and Neptune have lots more ice. Brrrrrr!!!

Right in the middle of the inner planets and the outer planets, we can find the asteroid belt. This belt is full of floating rocks, but there is more than rocks floating around. There is also metal but we don't know where it came from. Scientist thinks it comes from planets that were never formed millions of years ago.

Can you guess how many asteroids we have discovered so far? I will give you four choices:

1- 20,000,000 (Twenty million)

2- 3,250,000 (Three million two hundred and fifty thousand)

3- 150,000,000 (One hundred and fifty million)

4- 110,000,000 (One hundred and ten million)

Choose a number and find the correct answer on the next page...

Answer: The correct number is… 3!

Did you know there are over one hundred and fifty million asteroids out there? Yes! And each year we find more and more.

FUN PLANET FACTS FOR KIDS:

The solar system has amazing gas giants like Saturn. This planet is a lot like Jupiter but it has beautiful rings. Saturn's rings are very thin and made up of ice, water, and other materials. Saturn is also amazing because it has lots of satellites or moons. Do you know the name of Saturn's moons?

Here is just a few Pandora, Calypso, Helene, Rhea, Titan, Atlas, Ymir, Narvi , Hati, Prometheus, and Janud.

Have you ever heard of any of these moon names before?

Chapter 3: Mercury

Let's talk about each of the planets and see if we can learn anything interesting! Ready?

Mercury:

Mercury is near the sun. It is a hot planet and not a cool or breezy place to live. Sometimes, the temperature is so hot it can reach over 800 degrees Fahrenheit!

How hot is that?

The hottest temperature recorded on planet earth in the Libyan Desert was 136 degrees Fahrenheit. But guess what? Mercury is much, much hotter!

The other side of Mercury gets very cold. -300 degrees Fahrenheit. How cold is that?

The coldest temperature recorded Earth was -127 degrees Fahrenheit in Antarctica. Brrrrrrr…. But guess what? Mercury is much, much colder!

What else can we learn about Mercury?

This planet is small and has a lot less gravity than planet earth. How much less? ***Kidsastronomy.com*** says if you weigh 70 pounds on earth, you will only weigh 27 pounds on Mercury!

Mercury is called the morning star because you can see it before the sun gets up in the morning.

The days on this planet are long. On planet Earth our day is only 24 hours, but on Mercury it's 1,408 hours long or 58.6 earth days long. That's a long, long, day!

Chapter 4: Venus

Venus:

Venus and earth have lots in common. If you weighed 70 pounds on Earth you will weigh 63 pounds on Venus.

We do not know a lot about this planet, but we know it has a thick atmosphere with volcanoes, mountains, and sand.

Like Mercury, Venus is hot. It is even hotter than Mercury! You might wonder why this planet is hotter when it is further away from the sun. The secret is the atmosphere. It is so thick it traps the heat and keeps it inside the planet. Think of it like a big Greenhouse effect. If you do not know what this means, look it up online! Don't forget to get a parent's or guardian's permission.

Venus has temperatures of 864 degrees Fahrenheit! And when you get to the surface, it's a little hotter… 872 degrees Fahrenheit! Even the nights are around the same temperature.

What else can we learn about Venus?

This planet spins backwards, so the sun rises in the west and sets in the east. Venus has a small tilt of 3.39 degrees and does not change seasons.
It is impossible to live on Venus. Why? Mostly because the planet is mostly Carbon Dioxide and toxic for humans to breathe. It also has no liquid water to drink!

The days on Venus are much longer than on Mercury. It is 5,832 hours long or 243 earth days long. That's an extremely long day!

Chapter 5: Earth

Earth:

This planet is a pretty jewel in the solar system! It is full of life, good air to breathe, lots of water and great land to grow food.

The strong magnetic field in our planet helps to protect our atmosphere. And the tilt gives us beautiful seasons with lots of color!

DID YOU KNOW?

The earth rotates or spins on its axis. What's that? When the earth spins it takes one whole day or 23 hours, 56 minutes and 4 seconds!

Think of it like this: Stand in the middle of a room and spin around without moving out of place. Did you stand straight like a beanpole? Good job!

What about the earth? Does it "stand" straight? No, it doesn't! The earth **tilts at an angle**. Think of yourself as **leaning** forward (Without holding on to anything) or to the side without falling over! That's tilting! This is what the earth does. What's the angle? **23.5 degrees**.

This is why planet earth has such amazing seasons! If the earth did not spin at this angle, we would have the same season every day of the year.

What else can we learn about Earth?

Earth is called a water world because we are covered in more water than land: Approximately 71 percent! That's a lot of water. There is water in the air, glaciers, ice caps, and in the ground. This water helps our planet to stay cool and protects us from the burning sun. Earth is a wonderful place to live.

Chapter 6: Mars

Mars:

The red planet is fascinating. Why is Mars called the "red planet?" This is because Mars is covered in "rust-like" dust. Even the atmosphere is "pinkish red."

There are lots of canyons, channels, and plains. One of the largest shield volcanos is on Mars and called: Olympus Mons. NASA says this volcano is almost the same size as the state of Arizona, or 374 miles in diameter and 16 miles high!

Mars can get very cold at night. Around -120 degrees Celsius. If you are not sure how cold that is, here is a good conversion chart you can use to change Celsius to Fahrenheit from ***Kids.net.au.***

To convert Celsius to Fahrenheit: multiply the Celsius temperature by 1.8 and add 32 degrees.

F = 1.8 C + 32

To convert Fahrenheit to Celsius: subtract 32 degrees from the Fahrenheit temperature and divide the quantity by 1.8.

C = (F - 32) / 1.8

So multiply 120 x 1.8. The answer? **216**. Now add 216 + 32. The answer? **248**.

So 120 degrees Celsius is 248 degrees Fahrenheit!

Let's try the other way.

248-32 is equal to: **216**. Now divide 216 by 1.8. The result? **120**! Great work!

Chapter 7: Jupiter

Jupiter:

This planet is big with lots of color and gravity. If you weighed 70 pounds on Earth, you would weigh 185 pounds on Jupiter!

This planet is full of violent storms. Right now, there is a gigantic storm that has lasted for over 300 years! This huge red spot is called "the Eye of Jupiter," with super hurricane winds much stronger than planet earth.

This planet has three rings called Gossamer, Halo, and Main. Jupiter also has lots of moons and four of the biggest were discovered by Galileo according to ***Kidastronomy.com.*** They are called Io, Europa, Ganymede, and Callisto

What else can we learn about Jupiter?

Jupiter is so big that 1,300 earths can fit inside and it has a large ocean made of hydrogen and water. Dozens of moons circle this interesting planet, but a day on Jupiter is shorter than on planet Earth. It only takes 9.9 hours for Jupiter to complete one rotation.

Chapter 8: Saturn

Saturn:

This gas giant is big and light. Some say that if you get a huge bathtub and put Saturn in it...guess what would happen? Yes! It will float! Why? Because it's full of hydrogen and helium and this makes the planet very light.

Can you guess what size bathtub you will need? Do some research and find out.

What else can we learn about Saturn?

Saturn has lots of moons with horrible weather. The winds race around the planet at speeds of 800 kilometers per hour!

The rings on Saturn are beautiful with lots of different parts. There are 7 rings called: Ring A, B, C, D, E, F, and G.

Not every exciting names, but they are easy to remember.

What makes these rings so unique? They are made up of ice, dust, and rocks. Some are as big as a whole house!

The days on Saturn are hard to figure out, but astronomers say it around 10 hours and 47 minutes long.

Chapter 9: Uranus

Uranus:

This planet is very strange and is tilted to the side. This tilt is so strong that some seasons on Uranus last for more than 20 years! Sometimes the sun shines on one pole of the planet for more than 80 years. That's a very long time to have the sun shining in one place.

Uranus has a thick atmosphere made up of methane, helium, and hydrogen. There are lots of rings and moons around the planet.

Some of the moons have interesting names like Oberon, Ariel, Miranda, and Titania.

What else can we learn about Uranus?

This planet has a lot of pressure around it. This pressure is so strong astronomers believe there are millions of diamonds on or around the surface of the planet!

Chapter 10: Neptune

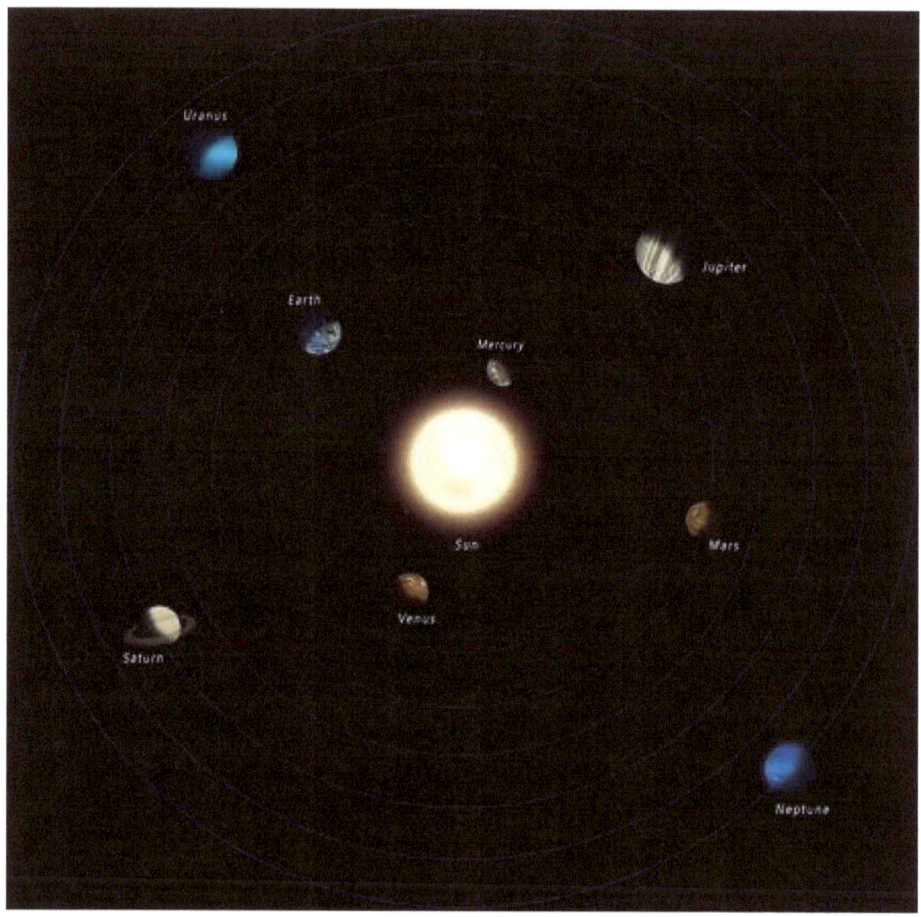

Neptune:

This planet was discovered many years ago by two men named Johann Galle and Heinrich D'Arrest. But even after this discovery we did not know a lot about the planet.

What we learned came in 1989 when Voyager 2 took lots of pictures of Neptune. This visit gave us a good idea of what this planet looks like in the solar system.

Neptune is a beautiful blue color with giant storms. It is windy and has five rings. One amazing thing is how strong the winds race around the planet. A few years ago, the "great dark spot" had winds of over 1,200 miles per hour!

Kidsastronomy.com says "No other planet in the Solar System has winds that are as strong as Neptune's."

What else can we learn about Neptune?

Neptune has 13 moons that we know. Here are some of their names Larissa, Proteus, Galatea, Nereid, Halimede, and Despina.

Chapter 11: Pluto

Pluto:

Pluto used to be part of our solar system, but not anymore.

On August 24, 2006, Pluto was named a dwarf planet. Because of this change, there is a new group of small planets called Plutoids.

Pluto is cold, rocky and light. If you weighed 70 pounds on Earth, you would weigh 4 pounds on Pluto!

What else can we learn about Pluto?

Pluto is a binary system with a large moon called Charon and four smaller moons named Nix, Hydra, Kerberos, and Styx.

The reason Pluto is not called a planet anymore is because it crosses Neptune's orbit. Instead of flat orbit, it is tilted. This means the planet orbits below and above the other planets in our solar system.

This planet is the only planet named by an 11-year-old child in Oxford. Her name was Venetia Burney. When Venetia told her grandfather about the name, he made the suggestion to the Lowell Observatory and it was accepted. It must have been very exciting for a young girl to name a planet!

Pluto can get very cold with temperatures from -235 to -210 degrees Celsius. Do you remember how to convert this reading to Fahrenheit?

What do you think about this amazing planet? Do you think it should be the ninth planet or not?

Share your answer with others and don't forget to get permission before you search!

Chapter 12: Interesting Facts

I hope you are enjoying this book on Beautiful Planets! Here are just a few more neat facts you may like to know.

- The Kuiper belt is a big group of dwarf planets, rocks, asteroids, ice, and dust that go around the sun. It is so huge it extends for millions and millions of miles after Neptune. It is on the outer part of our solar system, and Pluto is now a part of this group.

-The moon is more than just a pretty object in the sky. It keeps a good gravitational pull on the earth and keeps us steady in outer space. It causes the tides on the earth and helps them to flow in the right way.

- Our solar system is a very, very, big place. In 1977 Voyager 1 was sent from the earth to study the planets in our solar system. Guess what year Voyager 1 passed the last planet in our solar system?

1- 1935

2- 1990

3- 2011

The correct answer is number 2. In the year 1990, 13 years after it left Earth, Voyager 1 was saying goodbye to the last planet in our solar system. Amazing!

-Uranus takes a long time to go around the sun. Approximately 84 years. This means one of the poles is in darkness for 42 years and the other pole has daylight for 42 years.

-Jupiter is the biggest planet in our solar system. It is so big you could put all of the planets in the solar system inside of it and guess what? They would all fit!

-In 2005, astronomers found a new planetoid in our solar system. This celestial body is called Eris after the Greek goddess of strife. It takes approximately 550 years to circle the sun and it is bigger than Pluto!

-The new probe by NASA called New Horizons got to Pluto after travelling for 9.5 years. After it sent amazing pictures, it went off into space. Who knows what other amazing discoveries it will find!

Vocabulary:

Our solar system is full of amazing words you might like to learn about. Here is a small list to help you.

-Alpha Centauri	-Apogee	-Coriolis force
-Axial tilt	-Binary Star	-Cosmos
-Azimuth	-Bolometer	-Dark matter

- Geostationary
- Gibbous moon
- Hyper nova
- Inertia
- Interstellar
- Kilo parsec
- Zenith
- Meteoroid
- Mir
- Nebula
- Parallax
- Pulsar
- Quasar
- Red giant star
- Sidereal
- Singularity
- Umbra
- Van Allen Belt
- Wavelength

Do you know what these words mean? If you are not sure, ask your parent or a guardian's permission to search for the definition. I hope you learn something new!

Conclusion:

In conclusion: Our solar system is a very active place with lots of planets, asteroids, comets, meteors, rocks, ice and amazing dust clouds!

NASA even discovered a pink planet 57 light years from earth. And every year astronomers discover something new.

Did you know all of our fascinating planets can be seen with a telescope or binoculars? All you need to know is where to look!

Something else to think of!

Why don't you plan to use this information for show and tell, or another school project? Talk to your classmates about our solar system and share with your teachers. Maybe you can tell them why you like it so much, and what makes each planet, asteroid, comet, dust cloud ,or star unique!

You may choose to make this subject a science project or an experiment. If you do, don't forget the steps you need to make it a good science project.

Steps:

1 – You need to ask a **question** to be answered by observation or experimentation. Make it a very interesting question so your classmates and teachers will be excited to learn the answer!

For example: What's inside a black hole? What would it feel like if you could go inside?

2 – The next step is to state a **Hypothesis**. This is a big word but Sciencekidsathome.com explains it like this: *It is a tentative*

explanation for an observation, phenomenon, or scientific problem that can be tested by further investigation.

So your hypothesis is what you think the results of your project will be when your research is all done!

3 – Next thing to think about is: **Procedure.** This is very important. Procedure will help you to find the answer to your question and prove what you are trying to say.

4 – **Results**. You will need to show your results and all of the information you collected for your project.

5 – **Conclusion**. Finish up with what you learned and then answer the question you had in Step 1. If you can't answer the question, explain why the question cannot be answered.

I know you will have fun learning about the solar system and all its wonders!

If you don't like the ideas in this book, put on your thinking cap and come up with your own conclusions! I am sure you will do an amazing job!

We hope you have enjoyed this book on Beautiful Planets.

Happy Learning!

Sources:

- http://www.kidsastronomy.com/
- http://spaceplace.nasa.gov/days/en/
- http://www.spacekids.co.uk/solarsystem/
- http://www.sciencekids.co.nz/sciencefacts/planets.html
- http://coolcosmos.ipac.caltech.edu/
- http://www.space.com/16080-solar-system-planets.html
- http://www.livescience.com/32294-how-fast-does-earth-move.html
- http://easyscienceforkids.com/all-about-earth/
- http://www.enchantedlearning.com/wordlist/astronomy.shtml
- www.randroades.wcpss.net

Author Bio

K. Bennett loves to write for both children and adults. Many different subjects are interesting to research, but writing for children is special to her heart.

Her favorite pastimes include reading, traveling and discovering new things. Each of these activities helps to fuel her imagination and acts like a blank canvas waiting for more stories.

She is intrigued with fantasy elements like hidden worlds and faraway lands. And basically anything that gets her imagination soaring to new heights!

Her writing credits include children books online, short stories for online magazines, and novellas listed at Amazon.com

Our books are available at

1. Amazon.com
2. Barnes and Noble
3. Itunes
4. Kobo
5. Smashwords
6. Google Play Books

**Download Free Books!
http://MendonCottageBooks.com**

Publisher

JD-Biz Corp

P O Box 374

Mendon, Utah 84325

http://www.jd-biz.com/

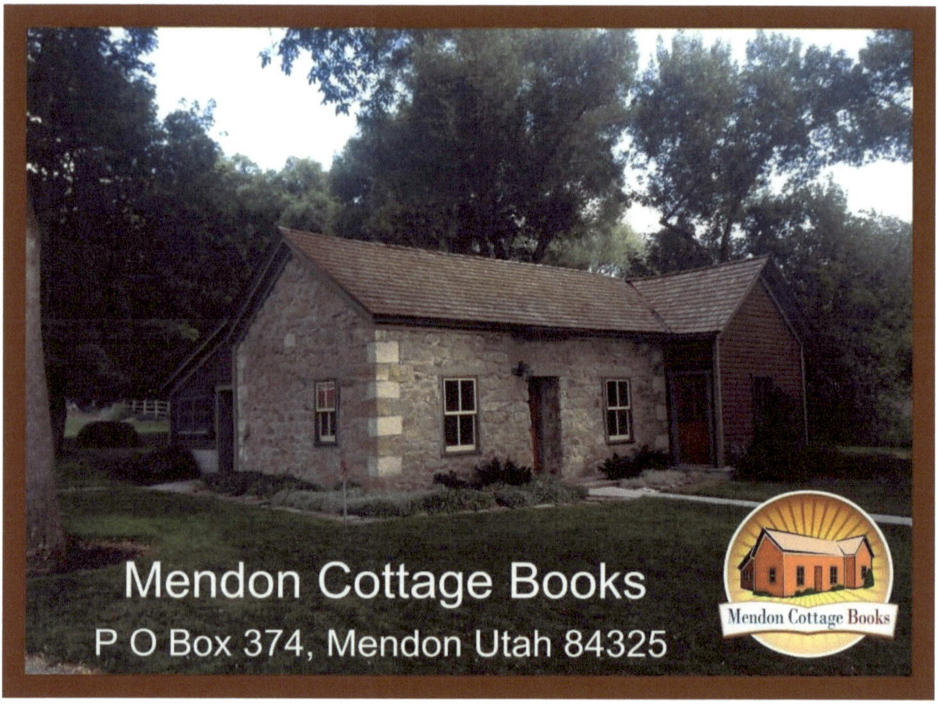

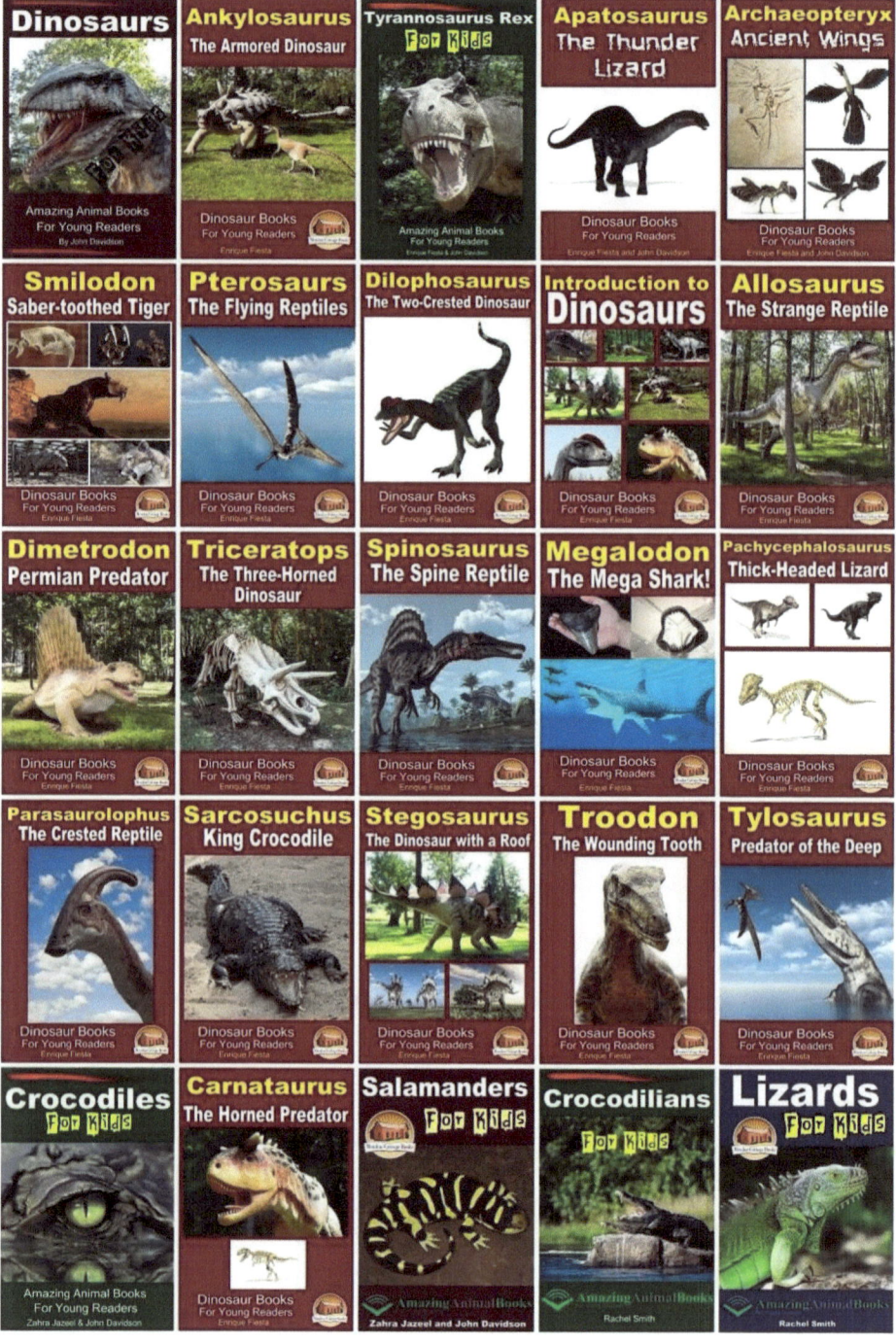

Beautiful Planets For Kids

Top Ten
Dog Breeds
For Kids
Amazing Animal Books
For Young Readers
Kisha Bennett & John Davidson

German Shepherds
Dog Books for Kids
K. Bennett

Bulldogs
Dog Books for Kids
K. Bennett

Dachshund
Dog Books for Kids
K. Bennett

Poodles
Dog Books for Kids
K. Bennett

Labrador
Retrievers
Dog Books for Kids
K. Bennett

Rottweilers
Dog Books for Kids
K. Bennett

Boxers
Dog Books for Kids
K. Bennett

Golden Retrievers
Dog Books for Kids
K. Bennett

Puppies
Dog Books For Kids
AmazingAnimalBooks
By John Davidson

Beagles
Dog Books for Kids
K. Bennett

Yorkshire Terriers
Dog Books for Kids
K. Bennett

Dogs
Top Ten Dog Breeds
For Kids
Amazing Animal Books
For Young Readers
Zahra Jazeel & John Davidson

Cats
For Kids
Amazing Animal Books
For Young Readers
K. Bennett & John Davidson

Foxes
For Kids
Amazing Animal Books
For Young Readers
Zahra Jazeel & John Davidson

Wolves
For Kids
Amazing Animal Books
For Young Readers
By John Davidson and Virginia Fidler

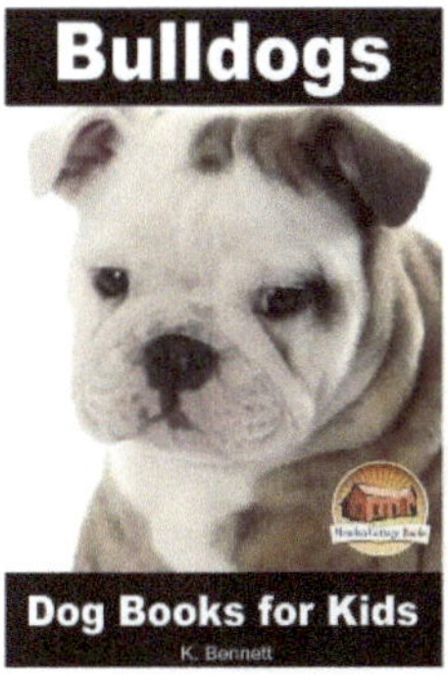

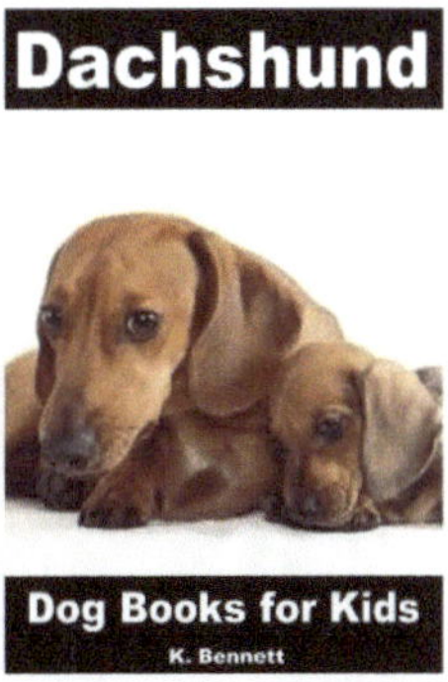

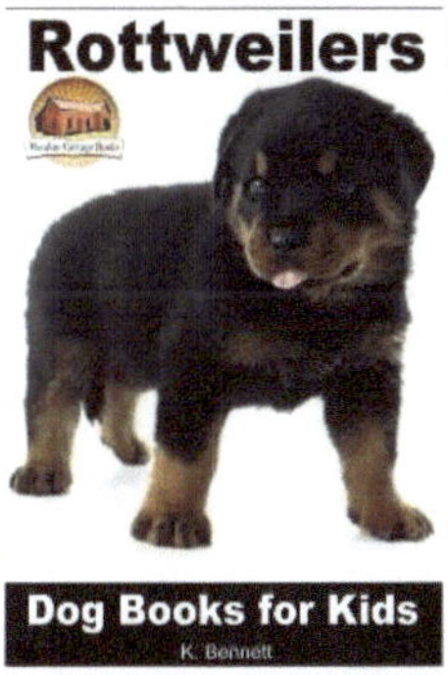

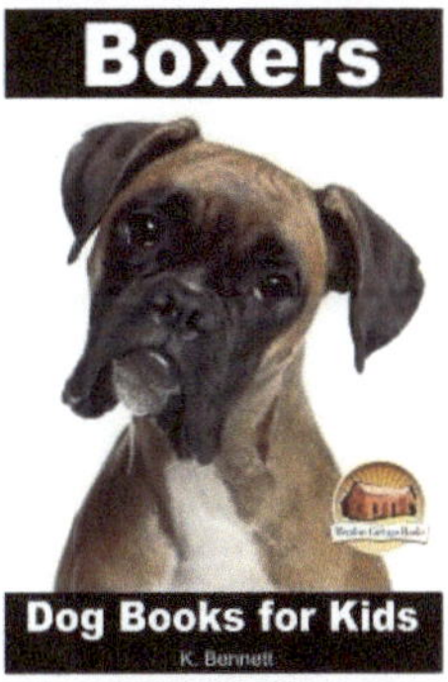

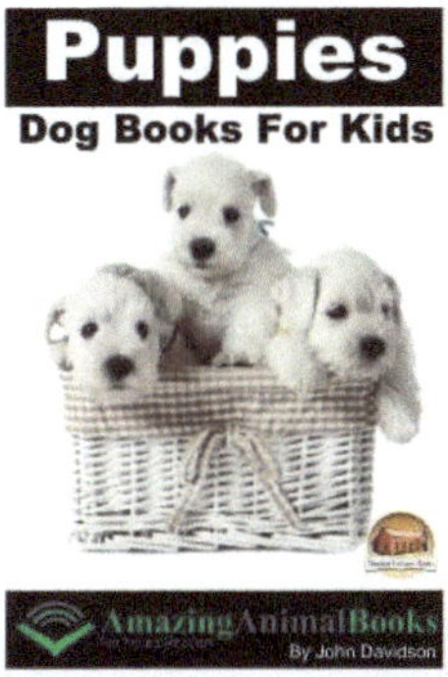

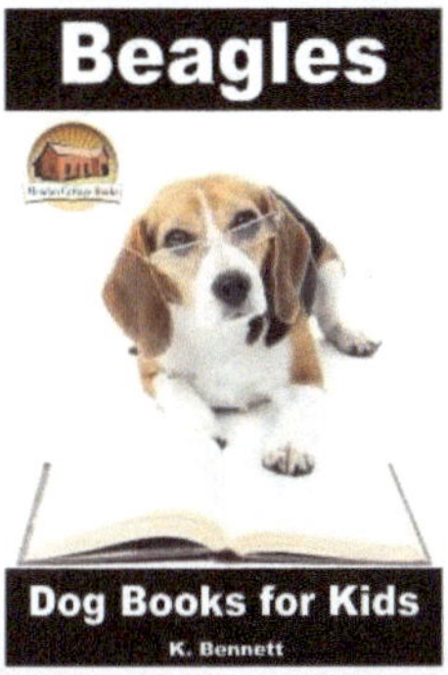

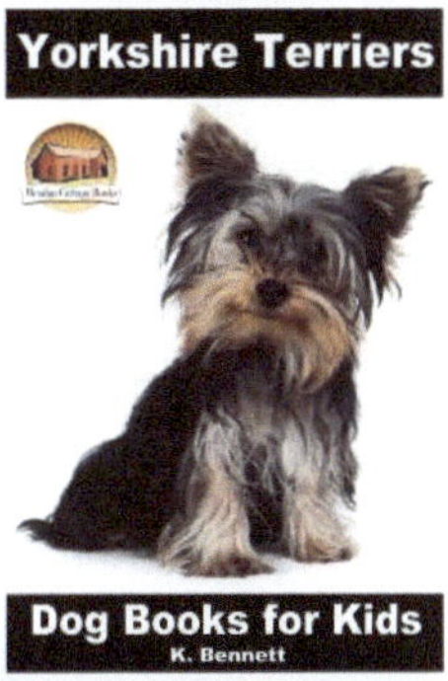

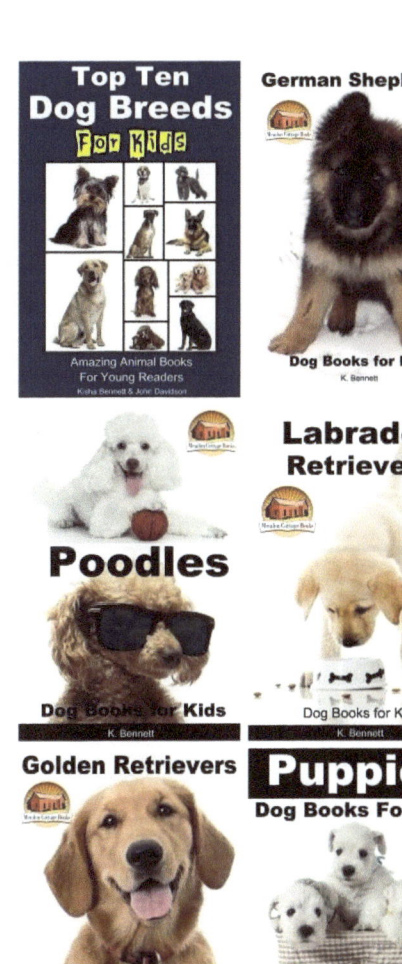

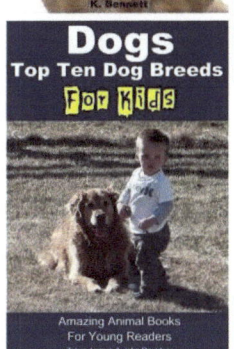

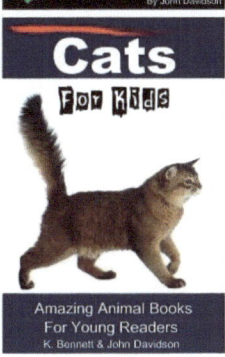

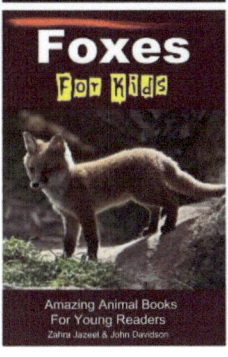

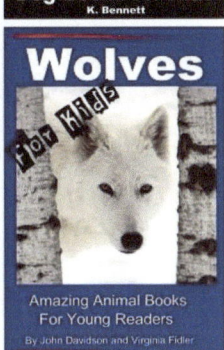

www.ingramcontent.com/pod-product-compliance
Lightning Source LLC
Chambersburg PA
CBHW041112180526
45172CB00001B/214